I0045539

James Freeman Clarke

How to Find the Stars

With Indications of the most Interesting Objects in the Starry Heavens

James Freeman Clarke

How to Find the Stars
With Indications of the most Interesting Objects in the Starry Heavens

ISBN/EAN: 9783337407575

Printed in Europe, USA, Canada, Australia, Japan

Cover: Foto ©berggeist007 / pixelio.de

More available books at **www.hansebooks.com**

HOW TO F

INDICATIONS OF TH

IN THE

AN AC

ASTRONOMICAL

JAMES

CONTENTS.

———◆◆◆———

CONTENTS

HOW TO FIND THE STARS.

§1.—OBJECT OF THIS BOOK.

THE object of this little book is to help the beginner to become better acquainted, in the easiest way, with the visible starry heavens; to know the winter and summer constellations, and the principal fixed stars. It will show the position of the constellations at different periods of the year, giving their place in each of the four seasons. It will also show how to find the separate clusters by a series of triangles and diagrams, covering the whole heavens, and connecting each constellation with its neighbors. It will indicate the most interesting objects at each period of the year, especially such as can be found with a telescope of moderate power. And it will close by describing the Astronomical Lantern, manufactured and sold by Lockwood, Brooks & Co., and its use.

§ 2.—THE FIXED STARS AND PLANETS.

The Planets (from the Greek word meaning " to wander "), which move among the fixed stars, cannot be represented on this Lantern. Nor is it necessary, since it is very easy to identify the principal planets : Venus, Jupiter, Mars, Saturn. Mercury is less easily perceived, but may sometimes be discovered just after sunset, in the western sky, especially in March. Seen through a telescope, even of low power, Venus, Jupiter, and Saturn are very interesting objects. Venus is often seen as a brilliant crescent, especially when it is brightest. I once showed it to a working man, through my telescope, and he manifested no surprise. He thought he was looking at the moon. Jupiter, with his belts and four satellites, always arranged on a nearly straight line, is surprisingly beautiful. Saturn, hanging in the sky, with its marvellous ring, self-suspended, surrounding it, is always an object of great interest. A telescope of low power shows these objects to advantage.

But though these planets are always interesting objects, they cannot be indicated on maps or globes, since they are continually changing their position. They may be distinguished from the fixed stars by the following characters : —

1. The stars shine with their own light, the planets with reflected light. Consequently, when a planet is nearly between the earth and the sun, it becomes a crescent; when between the earth and sun, and close to the sun, it disappears. A bright star may be seen, as bright as ever, when very near the sun.

2. The stars have no disks. A planet, seen through a telescope, shows a disk, which is larger in proportion to the magnifying power. Through a large telescope Jupiter appears round and as large as the full moon. But when examined by the same telescope, the brightest star, Sirius, for example, remains a point, with no apparent diameter, though its brilliancy is much increased.

3. The stars *scintillate*, or twinkle, — a phenomenon, the precise cause of which is not ascertained. But all the explanations of this phenomenon depend on the fact that the light of a star proceeds from a single point. The received theory of twinkling formerly was that the ray from the star was intercepted continually by motes floating in the atmosphere. But this theory ignored the fact that the mote, to produce this

effect, must always be as large as the pupil of the eye. The explanation of Arago is that this phenomenon is due to what is called the " interference of light," or light passing through atmospheric media not homogeneous.

4. The chief and most wonderful peculiarity of the fixed stars is that they virtually never change their positions in relation to each other. Thus the forms of the constellations remain unaltered for thousands of years. This indicates the enormous distance of the stars from the earth. Every year our globe, in its course round the sun, visits points, distant from each other in space more than 180,000,000 miles. Yet this makes no perceptible change in the relative position of the stars toward each other. This is what is meant by saying that the stars have no apparent annual *parallax.* By the most powerful telescopes and accurate micrometers, some slight annual movement is perceived in some stars; but to the naked eye no such movement is perceptible.

§ 3. — APPARENT MAGNITUDE OF THE STARS.

Though seen through the most powerful telescope, every star appears as a point of light, with no apparent diameter; yet some are much brighter than others, and are said, for convenience, to be of different magnitudes. Stars of the first six magnitudes alone are visible to the unaided vision; all below this size are telescopic stars.

As no instrument has been as yet brought into use by which to measure carefully the comparative brightness of the stars, the classification into magnitudes is quite imperfect. At present, it is as follows : —

There are about   20 stars of the First magnitude.
　"　　"　　"　　65　"　　"　Second　"
　"　　"　　"　　200　"　　"　Third　"
　"　　"　　"　　600　"　　"　Fourth　"
　"　　"　　" 1,100　"　　"　Fifth　"
　"　　"　　" 3,200　"　　"　Sixth　"

As the sizes of the stars diminish, the number increases.   Of telescopic stars, for example, there are of the

7th magnitude	13,000
8th magnitude	40,000
9th magnitude	142,000
And, visible in the most powerful telescope	20,000,000

At any one time and place, on a bright night, there are visible to the naked eye about 3,000 stars.

The names of the stars in the Northern Hemisphere, of the first magnitude, as usually given in the works on astronomy, are as follows: —

1. Sirius, in the Greater Dog.
2. Arcturus, in Bootes.
3. Vega, in the Lyre.
4. Capella, in Auriga.
5. Rigel, in Orion.
6. Aldebaran, in the Bull.
7. Betelguese, in Orion.
8. Antares, in the Scorpion.
9. Procyon, in the Lesser Dog.
10. Altair, in the Eagle.
11. Spica, in the Virgin.
12. Regulus, in the Lion.
13. Fomalhaut, in the Southern Fish.
14. Castor, in the Twins.[1]

The stars of the first magnitude, not visible in the Northern Hemisphere, are *Alpha* and *Beta*

[1] Burritt makes Castor of the 1st magnitude, and Pollux of the 2d magnitude. Rolfe and Gillet make Pollux of the 1st, and Castor of the 2d. So does Littrow (Himmel's Atlas). But Proctor (New Star Atlas) gives both as of the same size. This illustrates the arbitrary nature of this classification.

Centauri, *Alpha* and *Beta* Crucis, *Alpha* and *Beta* Argus, and *Alpha* Eridani.

§ 4.—POSITION OF THE CONSTELLATIONS AT DIFFERENT SEASONS OF THE YEAR.

By the revolution of the earth, which turns on its axis every twenty-four hours, every observer is brought round once a day to every point of the heavens, except that portion which is hidden by the rotundity of the earth itself. To an observer in the Northern Hemisphere all parts of the heavens become visible every year, except the region around the Southern Pole. Were it not for the daylight, which eclipses a portion, we should see all the other stars once in every twenty-four hours. But, during the year, we have an opportunity, by the revolution of the earth in its orbit, to see them all, except those at the extreme south.

January.

If we go out on a bright night in January, about nine o'clock, we shall observe the constellations in the following positions : —

The bright star nearly overhead, to the east of the zenith, is Capella. The constellation overhead, just west of the zenith, is Perseus. Lower down, to the west, is Andromeda ; lower still, the Great Square of Pegasus.

The Great Bear is in the northeast, standing on its tail. The Dragon makes a curve to the north, around the Little Bear. Vega, the bright star in the Lyre, is just setting in the northwest, with the Great Cross of the Swan standing upright. The long arm of the Cross points to the zenith, and half-way up in the northwest is Cassiopeia.

Towards the southwest, above the horizon, is Cetus, the Whale, a large constellation resembling an invalid's chair, its long back leaning toward Orion.

Orion stands nearly upright in the southern sky, with Sirius sparkling in the southeast. Below Capella, toward the east, are the Twins, two stars of nearly the same brilliancy. Procyon, the Little Dog, makes an equilateral triangle with Sirius and the bright star Betelguese on the shoulder of Orion.

The Sickle, in the Lion, is just rising in the east.

March.

At the beginning of March the following changes have taken place : —

Pegasus, Cygnus, and Cetus have disappeared in the west. Andromeda is low down in the northwest. Vega is just above the northern horizon, rising toward the east. Orion has moved toward the southwest. Aldebaran and the Pleiades are in the western sky, and above them hangs the bright Capella. Northwest of Capella is the W of Cassiopeia.

The Great Bear has moved up toward the zenith, his tail pointing east. Following its curve we see the bright star, Arcturus, rising in the east.

The Sickle, with the bright star Regulus in its handle, has moved up high in the southeastern heavens. Below it Cor Hydræ (the Solitary One, as the Arabs named it) is by itself, half-way between the south and the southeast.

To the southeast is Alkes, the chief star in the Cup.

The five principal stars of Virgo, which are shaped like a large cup, are rising in the east, below the triangle in Leo, of which Denebola is the apex.

The two bright stars in the head of the Dragon are just above Vega in the north.

June.

About the first of June, the following constellations have disappeared : Orion, Sirius in Canis Major, Taurus, Aries, Andromeda. Leo, with the Sickle, is sinking toward the west. Gemini and Perseus, with Auriga, have moved down to the northwest. Virgo is high in the south; Spica, its brightest star, is luminous in the southern heavens.

Below Virgo is the little quadrilateral of Corvus. In the south, just above the horizon, is the upper part of the constellation of Centaur.

In the southeast is the magnificent constellation Scorpio, the brightest of the southern groups, and, in advance of it, moving west, is Libra, known by three principal stars.

To the east of the Scorpion is the great constellation Ophiuchus. A line of six bright stars, three of which belong to Ophiuchus and three to the Serpent, rise from the horizon toward Arcturus, and then the line bends over to the northeast, imitating the form of a sabre.

Above Ophiuchus are Hercules and Bootes; the last star in the tail of the Great Bear being now near the zenith.

September.

The Great Bear has moved round so as to be between the Pole star and the northern horizon.

Arcturus and Bootes are now sinking toward the west. Virgo, Corvus, Hydra, the Lion and Sickle, the Twins, have disappeared. Capella, having gone round the sky, is rising again in the east.

Overhead is Lyra, and the Cross of the Swan. Below this, to the south, is the Eagle. Lower down in the south are Sagittarius, Capricornus, and Aquarius.

The Great Square of Pegasus is high up in the southeast. In the east Andromeda has risen into a conspicuous position, with Aries below. Farther north is Perseus, between Andromeda and Capella. Ophiuchus, the Serpent, and Hercules are all high up in the southwest.

When these movements have been watched during a year, they will always remain familiar and friendly. The stars, it will be noticed, revolve round the Pole star, in the opposite direction to the hands of a watch.

§ 5. — TO FIND THE STARS AND CONSTELLATIONS BY ALIGNMENT OR TRIANGULATION.

The constellations shown by the Astronomical Lantern, on its different maps, are these : —

1. Little Bear and Dragon.
2. Great Bear.
3. Cassiopeia and Cepheus.
4. Andromeda, Pegasus, Triangle, and Aries.
5. Perseus, and Andromeda, Triangle, and Cassiopeia.
6. Bootes, Hercules, and Northern Crown.
7. The Lyre, the Swan, and the Dolphin.
8. Orion and Taurus.
9. The Lions, Crab, Lynx, and the head of Hydra.
10. Eagle, Dolphin, and part of Pegasus.
11. Eridanus and the Whale.
12. Ophiuchus and the Serpent.
13. The Scorpion, Sagittarius, and Libra.
14. Virgo, Corvus, and Crater.
15. Auriga and Gemini.
16. Canis Major, Lepus, part of Argo and of Hydra.
17. Aquarius and Capricornus.

To find the position of these constellations is sometimes a little difficult to a beginner. The easiest way of studying the heavens so as to find the different star-clusters is by *alignment*, that is, by connecting the unknown constellations with those already known, by drawing lines between them.

The Great Bear.

We will begin with the two Bears, because these are known to almost all persons by means of the group called " The Dipper," from its resemblance to that common household implement. This consists of four stars making the cup, and three others the handle. Of these seven stars six are of the second magnitude, and one of the third.

The Little Bear.

A straight line drawn through the two outside stars of the Dipper nearly touches the Pole star, which is the star in the end of the tail of the Little Bear, or in the end of the handle of the Little Dipper. These two figures, the great and small Dippers, resemble each other very exactly, except that the handles bend in the opposite direction, and point to opposite regions of the heavens.

The Little Bear, attached to the Pole by the end of the tail, turns around it like the hands of a watch, but in the opposite direction. All the stars turn round the Pole every twenty-four hours, in the same direction.

Astronomers name the chief stars in a constellation by the letters of the Greek alphabet, *Alpha, Beta, Gamma*, etc. After these are exhausted they have recourse to the Roman letters, and then to figures.

As the seven stars in the Great Bear have often to be referred to in finding the others, their names are here given.

The two stars which point always to the Pole star are called " The Poinctrs."

The pointer nearest the pole is .	Alpha.	α
The other pointer is . . .	Beta.	β
The other star in the bottom of the cup is	Gamma.	γ
The other in the top of the cup is .	Delta.	δ
Of the three stars in the handle that nearest the cup is . . .	Epsilon.	ϵ
The middle of the three is . .	Zeta.	ζ
The end of the handle is . .	Eta.	η

Suspended from the two lower stars of the Dipper are two tassels, each consisting of three stars, two near each other at the end of the tassel, and the third nearer the Dipper. Another pair of small stars makes a third tassel.

Cassiopeia.

The Pole star is on the middle point of a straight line connecting the star *Delta*, of the Great Bear (where the handle joins the Dipper), with *Beta* Cassiopeiæ. This constellation consists of five stars of the third magnitude, in the form of a W. The star named *Beta* is at the right-hand extreme point of this W.

Cepheus.

A line drawn through *Alpha* and *Beta* Cassiopeiæ (that is, a prolongation of the right-hand line of the W) will strike *Alpha* Cephei. The principal stars of this constellation are three (*Alpha, Beta, Gamma*), making a curve between Cassiopeia and the Little Bear. They are all of the third magnitude. A line drawn through *Alpha* and *Beta* will strike the Pole star.

The Dragon.

A line drawn from *Delta* Cassiopeiæ (the star at the left-hand bottom point of the W), and passing through *Beta* Cephei (the middle star of the curve), will strike the head of the Dragon. This is as far from *Beta* Cephei as that is from *Delta* Cassiopeiæ. The Dragon's head is an equilateral triangle, formed by two stars of the second and one of the third magnitude. The other seven larger stars of this constellation (of the third and fourth magnitude) form a long curved line, bending round between the Great and Little Bear as far as *Beta* Cephei.

Andromeda.

A line drawn from the Pole star through *Beta* Cassiopeiæ, and prolonged as much further, will touch *Alpha* in the head of Andromeda. This star (*Alpha* Andromedæ) constitutes one of the corners of the great square of Pegasus. Andromeda contains five stars of the second and third magnitude, and some smaller ones. *Alpha, Delta, Beta, Gamma,* are nearly in a straight line.

Perseus.

A line drawn through these two last stars (*Beta* and *Gamma* Andromedæ) goes to Algenib (*Alpha* Persei). This is a star of the second magnitude, the largest one in a small curve of

stars, which, when once observed, is easily remembered. The remarkable variable star Algol
(the Demon) is the nearest bright star south of
Algenib.

Auriga.

A line drawn from the Pole star to *Alpha* Cassiopeiæ will make the base of an equilateral triangle, of which Capella, a very brilliant star of the
first magnitude, makes the third angle. The
three other chief stars of this constellation make
a straight line, pointing toward the North Pole.
A line drawn through the two stars making the
top of the Dipper (*Delta* and *Alpha*), and continued across the heavens in a direction *from* the
handle, will nearly strike Capella.

Taurus (the Bull), The Pleiades.

An isosceles triangle is formed by the three
stars *Epsilon* Cassiopeiæ (on the extreme left of
the W), Capella, and Aldebaran, a bright red
star in the forehead of the Bull. The group of
five stars to which Aldebaran belongs has the form
of a V, and is called the Hyades. Near the Hyades another group, smaller, clear and compact,
strikes the eye. These are the Pleiades.

Aries (the Ram).

A straight line from *Epsilon* Cassiopeiæ, through *Gamma* Andromedæ, being prolonged, passes through *Alpha* Arietis and *Alpha* in the Fish. This line is divided by these stars into four equal parts.

The Whale.

Alpha Arietis, *Alpha* in the Fish, and *Alpha* in the Whale, form an equilateral triangle. This is also the case with *Alpha* in the Whale, *Alpha* in the Fish, and the Pleiades. These four form, therefore, a nearly regular rhomboid. The seven principal stars in the Whale resemble a chair for reclining backward; and the constellation might be called " The Easy Chair."

The Twins (Gemini).

A line drawn from Capella through *Beta* Aurigæ (the next brightest star in the constellation) will pass not far from Castor. These twin stars, Castor and Pollux, are nearly of the same size, about four and a half degrees apart, that is, as far apart as the two stars in the Dipper which are nearest to each other. Castor and Pollux are both stars of the second magnitude, though Castor is a little the brightest. The other principal stars in this constellation are toward Orion and Taurus, and indicate the outline of a chair, leaning back, with the twins above, as a canopy.

Orion.

This, the most conspicuous constellation in the winter sky, is so well known that it does not need to be described. It contains two stars of the first magnitude, and three of the second. A line drawn through the two stars at the tips of the Bull's horns will strike Betelguese (*Alpha* Orionis), the very bright star on the northeast corner of the constellation. A line drawn from Castor, through Betelguese, will touch Rigel, which is the bright star on the southwest corner of Orion.

Great and Small Dog.

The three stars Pollux, Procyon, and *Alpha* Orionis make a nearly right-angled triangle. Procyon is a reddish star of the first magnitude, south of Pollux, in the Little Dog, and is at the right angle of this triangle. *Alpha* Orionis, Procyon, and Sirius make an equilateral triangle, constituted of three stars of the first magnitude, of which Sirius, the Dog Star, is the brightest in the whole heavens.

Leo, Hydra, and Cancer.

Castor, Procyon, and Regulus (in Leo) make another large right-angled triangle, turned in the opposite direction to the one last mentioned. In this, Procyon is again at the right angle. The five stars, Betelguese, Sirius, Procyon, Cor-Hydræ, and Regulus, make a large W, composed of three equilateral triangles. These five stars constitute the extreme points of the lines in this W, in the order named, beginning on the right upper point with Betelguese. Alfard, or Cor-Hydræ, is called the Solitary One, because it is the only large star in that region of the heavens.

About half-way between Castor and Regulus is the constellation of Cancer, or the Crab.

The chief stars in Leo make two groups. One of these is the Sickle, in the hand of which is Regulus. The form is so exactly that of a sickle, that when once seen it will not be forgotten. The other is a right-angled triangle to the east of the Sickle, having a star of the second magnitude (Denebola) at its most acute angle.

Bootes, Virgo, and Libra.

Arcturus, the chief star in Bootes, and one of the most brilliant in the sky, is easily found by continuing the curve of the handle of the Dipper. The first large star this curve encounters is Arcturus ; continue the curve further and it strikes Spica Virginis, another star of the first magnitude.

Libra follows Virgo from the east, and has three principal stars of the second and third magnitudes.

Northern Crown, Hercules, and Ophiuchus.

A straight line drawn through *Epsilon* and *Eta*, in the tail of the Great Bear (the first and last stars in the handle of the Dipper), will strike the Northern Crown. This is a small curve, composed of several stars, the concave side of which is turned toward the Pole.

The same line, continued, strikes the head and neck of the Serpent of Ophiuchus. Three stars of the Serpent and three in Ophiuchus appear standing up in the southeast sky, in the spring, in the form of a sabre.

A straight line from *Gamma* and *Epsilon* of the Dipper, strikes the Quadrangle in Hercules.

Lyre, Swan, and Eagle.

A line drawn through *Gamma* and *Delta*, in the Great Bear, goes through *Alpha* Lyræ (Vega or Wega). This is a splendid star, which rivals Capella in brilliancy. It makes a very striking isosceles triangle with Deneb in the Swan, and Altair in the Eagle,— all three being stars of the first magnitude. Two small stars make with Vega a beautiful little triangle. Four stars in the Swan, together with Deneb, make a conspicuous cross. Two stars of the third magnitude, on each side of Altair, constitute its attendants.

Pegasus, the Dolphin, Aquarius, and Scorpio.

A line drawn through Vega and Deneb intersects the Great Square of Pegasus. Between this square and Altair is the small constellation of the Dolphin. A line from *Alpha* Andromedæ, through the opposite corner of the Square of Pegasus, meets two stars near each other, in Aquarius.

The Scorpion, the chief summer constellation, is low down in the southern horizon; but the form is very distinctly perceived, and the bright-red star Antares, called the Heart of the Scorpion, is very brilliant in the summer sky.

Sagittarius, Crater, Corvus.

Sagittarius follows Scorpio from the east, its five chief stars resembling a bow and arrow, the arrow directed toward the Scorpion.

Corvus is a small quadrangle of four stars, easily recognized, to the south of Virgo.

Crater is another small constellation to the west of Corvus. Its three chief stars form a small equilateral triangle.

The Centaur is seen low down in the southern sky in May and June. It precedes Scorpio, and contains seven or eight stars of the third magnitude.

Fomalhaut is a star of the first magnitude, seen near the southern horizon, in October, below Aquarius.

Lepus is a small quadrangle of four stars of the third magnitude, below Orion.

If the line connecting Betelguese with Sirius be extended as much farther, it will strike a large star in the Ship. If the line connecting Procyon with Sirius be also extended as much farther, it will strike another large star in the Dove. These two stars, with Sirius, will then form another large equilateral triangle, almost exactly similar and equal to that formed by Betelguese, Procyon, and Sirius.

§ 6. — INTERESTING OBJECTS IN THE HEAVENS AT EACH SEASON OF THE YEAR.

Most of the objects to be described in this section are telescopic; but not requiring an instrument of high powers. In fact, for common observation a telescope of two or three inches aperture,[1] and three or four feet focal distance, is more convenient than a large one. Such a telescope, with powers varying from 25 to 150, will show many of the most interesting objects in the heavens. It will give a beautiful view of the moon, in all its phases, showing its craters and mountains; it will define the satellites of Jupiter, and the rings of Saturn; it will separate many of the double stars, and show their colors; it will reveal the nebulæ, and some of the fine groups or clusters of stars. Such a glass is more easily managed than a large one, and is much the best for those commencing the study.

Of groups of stars, some of the principal are the Pleiades, the Hyades, Berenice's Hair, the cluster in Perseus, and Præsepe, or the Beehive,

---

[1] This means the diameter of the object-glass at the end of the telescope nearest the object. The object-glass is the most important part of the instrument.

in Cancer. The last appears as a nebula to the naked eye. Few persons can see more than six stars in the Pleiades, when looked at directly; but by turning the eye sideways, we discover more. Miss Airy, daughter of the Astronomer Royal, England, has been able to count twelve.

The cluster called Berenice's Hair is midway between the star called Cor Caroli and Denebola in the Lion's Tail. It consists chiefly of a large number of small stars.

Præsepe, which seems like a nebula to the naked eye, is easily resolved by a telescope of small power into a collection of stars.

As the leading stars in each constellation are marked by Greek letters on all the maps and globes, we will give the forms and names of these letters here : —

a, Alpha.	ι, Iota.	ρ, Rho.
β, Beta.	κ, Kappa.	σ, Sigma.
γ, Gamma.	λ, Lambda.	τ, Tau.
δ, Delta.	μ, Mu.	υ, Upsilon.
ϵ, Epsilon.	ν, Nu.	φ, Phi.
ζ, Zeta.	ξ, Xi.	χ, Chi.
η, Eta.	o, Omicron.	ψ, Psi.
θ, Theta.	π, Pi.	ω, Omega.

Arabic names are also given to the largest and most conspicuous stars.

Circumpolar Constellations. These five constellations, Ursa Minor and Major, Cepheus, Draco, and Cassiopeia, never set to observers in most parts of the United States. The objects they contain can therefore be studied at almost any period of the year.

Ursa Minor, the Lesser Bear, has the Pole star at the extremity of its tail, and it swings round the Pole every twenty-four hours. The Pole star is 1° 23′ from the true pole, and in A. D. 2095 will be within less than half a degree of the polar point. The Pole star is a double star, with a small companion which can be seen by a power of 80 and an aperture of two inches, and is a good test for small telescopes. These two stars are of the second and ninth magnitudes.

Ursa Major, the Greater Bear, is said, by Cotton Mather, to have been so called by the Indians, long before they had any communication with the Europeans.

The names of the chief stars in Ursa Major are given in the directions for finding the constellations.

The middle star in the tail, Mizar, or *Zeta,*

has a companion, Alcor, which is visible to the naked eye. These two stars are $5'$ $11\frac{1}{2}''$ apart, and so form a good standard for measuring short distances. Another companion, of the fifth magnitude, may be seen by a telescope.

Alpha (named Dubhe), the Pointer nearest the Pole, has also a telescopic companion, of the eighth magnitude, of a violet color.

On the first day of January, at nine in the evening, the Dipper is in the northeast quarter of the heavens, with the handle pointing downward to the horizon. April 1st, at the same hour, it is nearly overhead, the handle pointing to the east. July 1st, it has gone round to the west of the Pole, the cup being now downward, and the handle pointing upward. October 1st, it is between the Pole and the northern horizon, the handle pointing to the west.

Beta and *Gamma*, the two stars in the bottom of the Dipper, are called the " Guardians of the Pole," as they march round it continually, like soldiers guarding a tent.

Alpha (Dubhe), the Pointer, is $28\frac{3}{4}°$ from the North Pole.

Alpha and *Delta*, the two upper stars in the cup, are 10° apart.

The Guardians of the Pole are 8° apart.

Benetnasch (Eta), the end of the handle, is of the second magnitude, and is 7° from Mizar (*Zeta*), in the middle of the handle.

Alioth (Epsilon), the third star in the handle, is 4½° from Mizar.

Megrez (Delta) is the smallest star of the seven, and is in a line with the Pole star and *Caph* (or *Beta*) Cassiopeiæ, which is equidistant from the pole on the opposite side. Both these stars are on the equinoctial colure.

Three pairs of small stars, one at the extremity of the right fore paw of the Bear, and two at the extremity of each of the hind paws, and all in a line, which line is nearly parallel with the Guardians of the Pole, indicate the limits of the constellation.

Cepheus. *Beta* in Cepheus is a double star of the third and eighth magnitudes.

Delta Cephei is a variable star, with a period of five days, eight hours, and forty-seven minutes. Its range is from the third to the fifth magnitude.

Cassiopeia. The star *Eta* is an easy double of the fourth and seventh magnitudes. Colors yellow and purple.

There is a loose cluster of stars, half-way from *Gamma* to *Kappa*, and another cluster, six degrees northwest of Caph (*Beta* C.).

Winter Constellations.

Some remarkable objects in the winter constellations are as follows : —

Orion. The star *Delta*, the upper star in the belt in January, is a wide double, of the second and seventh magnitudes.

Alpha Orionis (Betelguese) is a remarkable variable. It is sometimes brighter than Capella ; then it falls back until only a little brighter than stars of the second magnitude. Its period is about one hundred and ninety-six days.

Beta Orionis (Rigel) is a celebrated double star of the first and seventh magnitudes. It is a test for a three-inch aperture.

Zeta, the lowest star in the belt, is a close double.

The three stars in the belt of Orion make a line just three degrees in length, divided exactly in the middle by the central star, and so make a good measure of distances in the heavens. Hence these three stars have been called " The Yardstick."

The great nebula in the sword of Orion is one of the most extraordinary objects of this kind in the heavens. It appears, in a small telescope, as a mass of light cloud shining in the sky, with a singular black opening in the midst. In this opening are four stars, constituting what is called the trapezium. This nebula is not only one of the finest, but one of the most conveniently situated for observation.

Above this nebula is a cluster of stars, and also a multiple star, *Sigma* Orionis. The middle star of the belt has a distant blue companion, and is, itself, a nebulous star.

Below Orion is Lepus (The Hare), containing the following double stars : —

Xi Leporis (ξ), white and scarlet.

Gamma L. (γ), yellow and garnet.

Iota L. (ι), white and pale violet.

The star (α Monocerotis) in the left fore foot is a fine triple, but will appear as a double to a small power. *Delta* M. (δ) is an easy double.

Castor (which is east of the zenith in January) is a fine double star, easily separated. The components are about five seconds apart, both white. Castor and Pollux are about five degrees apart.

The Pleiades (easily found) are a beautiful cluster in the telescope, which reveals great numbers of stars.

The Hyades, in the Bull, near the bright red star Aldebaran, is another beautiful cluster, more diffused than the Pleiades. The star Aldebaran has been carefully examined by the spectroscope, and is shown to contain hydrogen, sodium, magnesium, iron, antimony, mercury, calcium, bismuth, and tellurium.

The Crab Nebula (1 M.) is near ζ Tauri (*Zeta*), and appears oval in a small telescope.

Some other remarkable objects are to be seen in the heavens in January. For example :

Presæpe, in the constellation Cancer, is rising in the east, below the Twins. With Castor and Procyon it makes a nearly right-angled triangle, the right angle being at Præsepe. To the naked eye it appears as a nebulous speck ; but a telescope of moderate power easily resolves it into a cluster of minute stars, of which forty have been counted with such an instrument.

Another cluster is in the sword-handle of Perseus, but is better situated for observation in March or April, when it is lower down in the west. A line drawn from Alcyone, the brightest star in the Pleiades, passes through this cluster to the middle of Cassiopeia. This is one of the most brilliant telescopic objects in the heavens.

Low down in the southwest, in January, is the large constellation, "The Whale" (Cetus), the chief stars of which arrange themselves in the form of an invalid's chair, the back leaning toward Orion. The star in the middle of this back is called Mira, and is a remarkable variable. It passes from the brightness of a star of the second magnitude till it becomes invisible. Its period is three hundred and thirty-one days, eight hours, and four minutes. It remains in its greatest brightness for about a fortnight, being then nearly equal to a star of the second magnitude. It then decreases during three months, till it becomes invisible, in which state it continues during five months, when it reappears, and increases during its remaining period of three months. After having disappeared to the naked eye, it may sometimes, but not always, be traced to its lowest point as a star of the twelfth magnitude by a telescope of considerable power, but sometimes disappears entirely.

Spring Months — Noticeable Objects.

Double Stars. One of the finest is *Gamma* Andromedæ, of the third and fifth magnitudes. The colors are deep yellow and sea-green. Andromeda, in April, is low down in the northwest.

The double star Castor is in a good position for observation at this time.

Gamma Leonis, the brightest star in the blade of the Sickle, and above Regulus, in the south, is a fine double. The magnitudes are second and ninth.

Gamma Virginis, a " wonderful pair" of the fourth magnitude each — colors, silvery-white and pale-yellow. In 1836 they were so close together that no telescope could separate them ; but they now are so far apart that a telescope of moderate power shows them.

Sigma Coronæ Borealis — magnitudes six and seven ; distance three seconds ; colors, yellow and blue. The colors and sizes are very variously reported.

Clusters and Nebulæ. The nebula in Andromeda is one of the finest in the heavens. It is in the form of a lens, and nearly half a degree long ; visible to the naked eye on a clear night. This is one of the irresolvable nebulæ, which defies the power of the Cambridge refractor, which extended it to the surprising dimensions of four degrees of length.

Cluster in Hercules. This is perhaps the finest object of the kind in the heavens. It is a superb globular cluster, in which thousands of stars appear. It will be found between *Eta* and *Zeta* of Hercules. It is just visible to the eye.

Variable Star. The star Algol, in the head of Medusa, has the very short period of two days, twenty hours, and forty-eight minutes. During two days and fourteen hours it is of the second magnitude; and during the rest of the period it gradually diminishes to the fourth magnitude, and returns again to the second. Its name in Arabic means "The Demon."

Objects in Summer and Autumn.

Double Stars. Early in summer a fine quadruple can be seen in the northeast near the beautiful and brilliant star Vega in Lyra. The star *Epsilon* Lyræ, which makes with *Zeta* Lyræ and Vega a small equilateral triangle, is separated by a good telescope into two stars, each of which is a double star.

Zeta Lyræ is a splendid and easy double; colors topaz and green; magnitudes four and five.

Eta Lyræ is a wide double.

Alpha Herculis. Magnitudes three and five. Colors, orange and emerald green. Distance five seconds. "A lovely object, one of the finest in the heavens."

Kappa Herculis. Magnitudes five and seven. Distance thirty-one seconds. Color, yellow. An easy double.

Beta Scorpionis is a fine double. Magnitudes two and five. Distance thirteen seconds. Colors, white and lilac.

Xi Scorpionis is a double in the claw farthest east. Magnitudes four and seven. Distance, seven seconds.

39 Ophiuchi is a double in the right foot of Ophiuchus. Magnitudes five and seven. Distance, twelve seconds.

70 Ophiuchi, in the right shoulder. Colors, yellow and red. Magnitudes, four and a half and seven. Distance, five seconds.

67 Ophiuchi, near 70. Magnitudes four and eight. Distance, fifty-five seconds.

Epsilon Bootis. Colors, pale-orange and sea-green. Magnitudes, three and seven. Distance, two or three seconds. Called Pulcherrimum on account of its great beauty.

Beta Cygni (Albireo). One of the finest doubles in the heavens. Colors, yellow and sapphire-blue. Magnitudes, three and seven. Distance, thirty-four seconds.

Clusters and Nebulæ. A nebula in Lyra, the only annular nebula accessible in common telescopes. Easily found, one-third of the distance from *Beta* Lyræ toward *Gamma* Lyræ. Somewhat oval.

A nebula between *Alpha* and *Beta* Scorpionis. Described by Sir William Herschel as "the richest and most condensed mass of stars in the heavens." Like a comet.

A nebula in Sagittarius (M. 22) "is a valuable object for common telescopes on account of the visibility of its compounds." Draw an imaginary line from the upper end of the bow to the hand of the Archer, and this nebula will be found upon it. This cluster is on the Sun's path, and another cluster on the same circle, due west (M. 8), is a splendid galaxy, visible to the naked eye. A bright coarse triple star is followed by a luminous mass.

Five M. in Libra is "a beautiful assemblage of minute stars, greatly compressed in the centre."

We have given only a few of the conspicuous objects, and for the rest refer the observer to such work as Proctor's "Half-hours with the Telescope," and "Half-hours with the Stars;" Edwin Dunkin's "The Midnight Sky;" Webb's "Celestial Objects for Common Telescopes;" J. J. Von Littrow's "Atlas des gestirnten Himmels" (German text, and good maps, and cheap); "Chambers's Descriptive Astronomy" (8vo, 1867); Burritt's "Geography of the Heavens, and Atlas;" "Atlas of the Heavens in Six Maps," published by the Society for the Diffusion of Useful Knowledge; Dick's Astronomical Works, and those of Arago, Herschel, Nicol, Rolfe, Bouvier. Burritt's School Atlas is cheap and good, though the plates are now somewhat worn. Perhaps the best Star Atlases are those of Proctor ("A New Star Atlas," second edition, by R. A. Proctor), and of Argelander ("Uranometria Nova;" containing seventeen charts of the Heavens); together with that of Hind, in Keith Johnson's Atlas of Astronomy.

§ 7. — DESCRIPTION OF THE ASTRONOMI- CAL LANTERN AND THE SLIDES.

The object of the Astronomical Lantern is to facilitate the study of stellar astronomy. It is intended for beginners, for astronomical classes in the high schools or private schools, and, in fact, for all who desire to become acquainted with the constellations.

The difficulty hitherto experienced in this study, and which is obviated by the use of the Lantern, is this: In order to study the starry heavens, it has been necessary to use an astronomical atlas, or a celestial globe. These must be examined in the house, by the light of a lamp. The observer, having found his constellation on the atlas, goes out to look for it in the sky. But, by the time he gets out of doors, he has forgotten how it looked on the atlas. And when he has found it in the sky, he has forgotten how it looked there, before he gets back to his atlas or globe. All who have studied the constellations have met' with this difficulty.

Now, the Astronomical Lantern will make the study of the stars perfectly simple and easy. It is constructed like a dark-lantern, closed on three sides, and on the fourth provided with a ground glass, in front of which slides can be inserted. On each of these slides, which are semi-transparent, is represented a constellation, the places of the stars being indicated by perforations, through which the light shines. The largest perforations in these slides are for the stars of the first magnitude, and they are made smaller, in due proportion, for the lesser stars. The student, therefore, wishing to observe any particular constellation or cluster, has only to light a candle within the Lantern, insert the appropriate slide, and go out into the night. He holds up the Lantern in one hand, and can compare, at his leisure, the constellation as it appears on the Lantern with that in the sky, until he becomes perfectly familiar with the latter.

It is easy to see how much the use of such a Lantern facilitates the whole study. In fact, we think that henceforth no one wishing to become acquainted with the heavens can afford to dispense with it. The increased ease of the study will probably also enlarge the number of students in this interesting department of science. We all are glad to know the names and positions of the stars. For, though Shakespeare has said, —

"Those earthly godfathers of Heaven's lights,
 Who give a name to every wandering star,
 Have no more profit of their shining nights
 Than we who walk and know not what they are; "

yet it must be confessed that to recognize the famous stars and groups which have been referred to since the days of Job, in the literature of all nations, is no small satisfaction.

This invention was patented December, 1870, and is manufactured and sold by Lockwood, Brooks & Co., Boston.

To use the Lantern, it is necessary to see what constellations are favorably situated for observation at the time; which can be done by the help of this little treatise. A slide is then selected, containing one of the groups of stars which are above the horizon; it is then put in its place and the candle is lighted. Then, by holding it up, and comparing the slide thus illuminated with the stars themselves, the principal stars and their positions will be easily fixed in the memory.

The card-slides accompanying the lantern are seventeen in number, and contain all the con-

stellations visible to an observer in the Northern
Temperate Zone. Other slides can easily be
added as required. In these maps of the constellations we have retained the names and the
designations of the stars, but have omitted the
figures of Bears, Bulls, Unicorns, Sheep, Virgins,
Dragons, Lions, and the like, which have so long
disfigured the celestial globe. Instead of these
confusing figures, few of which bear any resemblance to the constellations, we have substituted
dotted lines, tying together in simple figures the
chief stars in each cluster. Experience shows
that by these diagrams the separate constellations
are much more easily recognized and remembered
than by the traditional pictures of animals,
monsters, and men, which have hitherto crowded
the starry atlas. By these connecting lines, the
principal stars in each group are easily found and
associated in the memory.

In preparing these slides we have followed the
"Uranometria Nova" of Argelander. (Trans-
lation of the title: "The New Uranometria. A
representation of the stars visible to the naked
eye in Central Europe, the magnitudes being
taken from immediate observation of the heavens. By D. Fr. Argelander, Professor of As-

tronomy, and Director of the Observatory at
Bonn. Berlin, 1843.") This atlas was selected
because of its reputation for accuracy, and because the scale by which it is drawn was best
adapted to the size of our slides. The stars
of the first four magnitudes are perforated by
means of punches of the appropriate size. At
the top of the map is given the names of the constellations which it contains. At the bottom is
given their position in the heavens at such time
of the year as is suitable for observation. The
stars are lettered with their proper symbol.
Double stars are indicated by a D. The nebulæ
are shown by means of a group of minute dots,
and star-clusters in a similar way. On each map
is also a list of the telescopic objects which are to
be found in the constellations represented upon
it,— those, at least, which are suitable for small
telescopes. In this way the lantern may be of
great use to observers possessing such instru-
ments, by enabling them to find easily the double
stars, clusters, etc., which are in a convenient
position for observation at any period of the year.
Those who have spent hours in looking through
books of astronomy, in order to see what suitable
subjects for their telescopes are above the horizon

at any particular time, will easily understand the advantage of this arrangement. In choosing among the various doubles and nebulæ, we have made much use of Mr. Proctor's valuable little work, "Half-hours with the Telescope;" and in locating these objects we have constantly consulted Mr. Proctor's "New Star Atlas," comparing it continually with the well-known maps published by the Society for the Diffusion of Useful Knowledge, both of which are very rich in nebulæ.

We will only say, in conclusion, that we hope this simple apparatus may facilitate the study of the science which offers so much pleasure and instruction to its votaries.